Cotton

A 50 Year PICTORIAL History
The Photographs of Harris Barnes

Published by:
True Exposures Publishing, Inc.
PO Box 5066
Brandon, MS 39047
601-829-1222
1-800-323-3398
www.trueexposures.com

Distributed by:
Rural Services
537 School Street
Clarksdale, Mississippi 38614
662-624-8986
662-627-2088 Fax

© Copyright 2002 by Harris Barnes

All rights reserved. No part of this publication may be reproduced,
stored in a retrieval system, transmitted, or used in any form or by any means,
electronic, mechanical, photocopying, recording or otherwise without
prior permission of the copyright holder.
The publisher is not responsible for the accuracy of
the information contained within this book.

Library of Congress Number: 2002106431

ISBN: 0-9642595-5-9
Printed in Korea

Publisher: Paul T. Brown
Publishing Assistant: Terry Mullen
Edited by: Harris Barnes
Graphic Production: Heckler Design
Scans: Service Printers, Inc.
All photography © Copyright Harris Barnes

DEDICATIONS

To my wife, Jamye seen here with me in 1942 on the steps of our "Honeymoon Quarters" on the US Marine base, Camp Lejeune, North Carolina. We celebrated our 60th Wedding Anniversary on June 13, 2002!

In loving memory of Jamye.

TABLE OF CONTENTS

LAND PREPARATION	14 - 19
PLANTING	20 - 27
SEEDLING COTTON	28 - 33
CULTIVATION	34 - 39
CROP ROTATION	40 - 47
HOES & HERBICIDES	48 - 53
IRRIGATION	54 - 60
SKIP ROW	61 - 66
STAGES OF COTTON	68 - 69
INSECT CONTROL	70 - 76
MATURE COTTON	77 - 81
OPEN COTTON FIELDS	82 - 88
DEFOLIATION	89 - 92
HAND/MECHANICAL PICKING	93 - 105
HUMAN INTEREST	106 - 107
MODULES	108 - 116
GINS	117 - 122
BALES /WINTER	123 - 128
PEOPLE	129 - 131
ILLUSTRATION DATA	132 - 142

ACKNOWLEDGEMENT

To my friend, Billy Connell, with some friends, who got me interested in photography in the late forties, so that I could record the growth of our four "baby boomers". This lead to a wonderful life in agricultural photography.

Capt. Harris Barnes, USMC, in combat gear in 1944 outside of San Diego, California prior to shipping out to the South Pacific.

PREFACE

Harris Barnes is the quintessential Southern Gentleman, a man of many cameras with the technical expertise to get exciting images from each, and a walking encyclopedia of cotton production. He ranks easily among the top ten farmers whose enthusiasm and energy brightened my four decades as a Farm Journal editor writing about cotton.

Joe Dan Boyd
Former Editor of *Farm Journal Cotton Today*

I am very pleased to acknowledge the very extensive coverage of agriculture by Harris Barnes over the past 50 years, as a Farm Manager from 1946 to 1969 and from 1970 to the present as a farm journalist and photographer.

Harris is a class act, as evidenced in the Marine Corps, his church and family life and in his profession. I have stated many times and will restate that were I President of the United States I would name Harris Barnes, Jr., as my Secretary of Agriculture.

David Mullens
Retired Cotton, Soybean, and Rice Farmer, Clarksdale, MS

This is a grand journal of Harris Barnes' long lived romance with God's good earth and His crop cotton.

Through his artistic eyes and skilled command of the camera, the author carries us on the varied and annual excursion men take when they dare turn the nettles into the gray, silty soil and seed cotton into that fertile womb.

He moves the reader from those days of yore when barefooted men in spring plowed the chilled earth, through the steamy days of summer and into the cool autumnal season when men, women and children on bended knee gathered the seed filled lint.

The octogenarian author shares his life with cotton in these colorfully and pithily filled pages. He teaches a lesson of the amazing progress man has made in the production and the harvesting of the internationally wanted lint.

His testament is singular, honest, vivid and enlightening. Stewards of the land today and into the distant tomorrow are the better for this sharing of time and talents by Harris Barnes of his adulthood affinity with cotton and her producers during the twentieth and the early days of the twenty-first centuries.

His is the vividly told story, in picture and words, of a marvelous plant and her people.

William H. Craven, Jr.
Retired County Agent
Waynesboro, Georgia

FOREWARD

I thank the Lord, almost without ceasing, for my wonderful family, my Christian parents, brothers and sisters, my wife, our four children and 10 grandchildren.

There are so many other things for which I am also thankful. My cup does indeed runneth over.

That the Lord permitted me to live through a war, which was supported by most Americans, ranks high on my attitude-of-gratitude list. That I was allowed to witness firsthand the "greatest generation" of farm chemicals, machinery and production practices, especially for cotton, is another chart-topper for me.

Thank you, Lord!

It all seems at once so long ago and as recent as yesterday.

There was summertime and the living was not so easy just before World War II on my dad's small farm. I helped him measure cotton fields and conduct a bit of agricultural cartography for those government maps of the "plow-up days" during the late 1930s and early 1940s.

The bottom line is that a kind of harmonic confluence of events combined to equip me with an unofficial hands-on education in the rural sociology of that day and time: life, work and conditions of the average Mid-South farm.

It would serve me in good stead when, just after VJ Day, my Marine Corps uniform went into the closet, and I became a Farm Manager. Surprisingly, in 1946, very little had changed during my wartime absence.

Tenant families were still chopping and picking their five-to-ten-acre crops. A happy marriage of mules, double shovels, International F-20s and John Deere A's was still the cultural norm. Available insecticides, for the most part, only "fattened" boll weevils and "nurtured" bollworms and bud worms.

Herbicides for cotton? Forget it. That was a dream yet to come true.

We did have the daring young pilots in their Stearman flying machines, Stearman pilots who "dusted" cotton plants with insecticides and defoliants of the day in powder form. And we had "Black Annie" a 16% nitrogen material in powder form which would drop the leaves from cotton plants as well, maybe better than present-day liquid defoliants. As I said, it was a long time ago, but that's the way I remember the good old days. They really weren't all that bad.

Tenant families still "shook out" their pick sacks, seven or nine feet in length, into their cotton pens. When a family had picked enough seed cotton—about 1,300 pounds to gin out a 500-pound bale—they raised a flag and the "day crop" folks would pull up to the shed in a wagon or trailer that held one or two bales to "load it out" with a woven white oak basket.

Our two-stand plantation gin could process two bales per hour. On a really good day we could "tie out" 25 bales in a single workday.

To me, it all began to "happen" when International Harvester marketed its first one-row picker in 1948. It had already been field proven on the Hopson Plantation near Clarksdale, Mississippi. Granted, it was not a pretty sight. The amazing new rig left seed cotton on the stalk, dropped cotton in the middles and missed entire stalks on the row ends. Yet it was the answer to the prayers of a cotton economy in the throes of change. Many of the better men and women were leaving the low-paying jobs of the South and moving to the North for higher wages.

Pre-emergence herbicides for cotton appeared on the scene at a most opportune time. In 1952 we were losing not only the hand pickers, but also the hand choppers.

The transition to herbicides was not without cost. Applications of Dinitro that year actually killed cotton on sandy soils. It was a lesson we would not forget. But applications of CIPC performed nobly - after activation - that same year.

Cotton was off to the races with the "wonderful world" of herbicides. It was an exciting time to be a cotton planter.

Many of the pre-plant-incorporated herbicides came on stream in the early-to-mid 1960s.

Thank you again, Lord!

Flame cultivation was a novel idea that also appealed to many cotton producers at about that time as well.

In contrast, it's hard for younger cotton folks to realize and fully appreciate the blessings of today. Cotton varieties bred to be tolerant to Buctril and Roundup for production in no-till systems, ultra-narrow-rows and stale seedbeds.

I remember 1957 as a banner year for cotton insecticides and defoliants and now can only marvel at today's technology that produces varieties less attractive to bollworms and bud worms because of a built-in Bt gene. The Good Lord must love the fabric of our lives.

Who can forget the early 1970s when we saw research developments, and the eventual introduction of the industry's first module builders? Who would have thought these seed cotton compactors would become about as important to the harvest process as the cotton gin and the mechanical picker?

And the cotton gins of today: One stand in any of the several multi-million-dollar gins can turn out as many bales in an hour as that old plantation gin could process in a day.

Since 1946 I have seen many "new, strange and exciting" innovations, machinery, chemicals and technology used in the production of cotton. I am grateful to have been a participant and a witness to all that.

But, more than anything else, I am so thankful that Billy Connell turned me on to photography more than a half century ago. Because of that I was motivated to record the production of cotton from the "good ole days" to the 21st Century in chronological order.

I particularly want to thank family and friends who helped me bring this pictorial history of production to a fruitful harvest and market. These include my son, Dudley McBee, my Financial Advisor, who was named for my best friend, Dudley McBee of Greenwood, Mississippi, who was killed while strafing a village that was held by the Japanese on Bougainville; my brother, W. A. "Bill" Barnes, retired Mississippi State Tax Commission; Joe Dan Boyd, Contributing Editor of *Farm Journal Cotton Today;* Nan Hughes, Pharr Brothers Advertising; T.M. Luster, retired Cotton Grower, and Cliff Porterfield, retired Coahoma Chemical Consultant, each helping with picture selection and technical assistance.

It's been a great ride and a fascinating read.

This book is my tribute to a life-long love affair with a plant that has always been much more than a crop. For me, and I suspect for you, cotton is also a culture.

We all are steeped in that culture.

HARRIS BARNES
Clarksdale, Mississippi
January, 2002

From the seed to the open boll, there are many practices, chemicals, and machines involved. In this book Harris Barnes has captured on film the most important - from the end of World War II to the new Millennium.

LAND PREPARATION

Left to Right: Four up pulling two-row disc harrow   Tenant plowing water furrows

Top Left to Right: Turning of Delta soil   Land being prepared by field cultivators on a large farm
Bottom Left to Right: Four-bottom mold board plow in Coastal North Carolina   Setting up four beds with disk hipper

Left: Four-row Do-All   *Right Top and Bottom:* Cliff Porterfield calibrates a broadcast spray applicator   Ground sprayer

Left Top and Bottom: A 12-row hipper   Running water furrows modern version   *Right:* Springtime-planting and turkey time

Left to Right: Six-row switch bottom plow in South Plains of Texas   Wheat planted in the middles to save the soil and seedlings

Left to Right: Ray Young, Pioneer of the stale seed bed   Big equipment working from "can to can't"

PLANTING

Top Left to Right: Two-row mule planter   A four-row "running board" planter   *Bottom Left to Right:* 1952 saw the first use of pre-emerge herbicides   In the early fifties, a pre-emergence herbicide "Ho-No-Mo" and seed in 100 % cotton bags

Left to Right: John Deere eight-row riding planter   Thurston Pellum, Farm Manager, checks depth of seed

Left to Right: Stovall Farms preserves the past   Planting cotton on the banks of the Mississippi River in West Tennessee

Left Top and Bottom: Planting as seen through the legs of a farm manager   Models Donna Moore Surholt (left and Mary Margaret Humber Polles in 100% cotton taking soil temps to determine the proper time to plant   *Right:* Sunset on Moon Lake

Left to Right: Planting seed in a stale seed bed in Louisiana   Cottonseed being dropped in wheat windbreaks in the Bootheel of Missouri
Opposite page: John Deere tractors with 12-row planters in Delta country

Top Left to Right: What EPA would like cotton growers to wear   A large modern planting operation
Bottom: No-tillage planting in West Tennessee

Top Left to Right: A spindle for growers to burn empty paper bags to prevent the land from being cluttered   The world's widest planter drops seed in Tennessee   *Bottom:* Essentials for planting and excellent stands - good men, machinery, seed and chemicals

SEEDLING COTTON

Healthy stands of cotton seedlings

Left to Right: Hill dropped cotton and cypress trees   A "Cabin in the Cotton" in the Tennessee Valley of North Alabama

Left: Old timers contend there is more cotton to be made per acre on crooked or curved, than on straight rows
Right: Four and one skip-row cotton

Top Left to Right: Beth Bobo shows how cotton should look in early June   Perfect stands of healthy cotton
Bottom: No-till cotton in the Tennessee Valley of North Alabama

Left to Right: Without wheat, a dust storm kicks-up in the Bootheel of Missouri   "Killed" wheat protects the seedlings

Left to Right: Wheat has saved Ron Lea's soil and cotton seedlings   Cotton seedlings protected by "killed wheat"

CULTIVATION

Left: Four-up pulls a two-row riding cultivator
Right: Anthony Harris cultivates his five-acre crop with a double shovel

Left to Right: Five tractors with umbrellas and four-row cultivators   A four-row cultivator

Left to Right: PTO type cultivator   Cotton grower supervises cultivation

Left to Right: A popular four-row cultivator   A Bell type post-directed herbicide applicator

Left to Right: Service truck of the mid-sixties
Two and one skip-row cultivators (axles of tractors not long enough to run both wheels between the cotton rows)

Top Left: 10-row cultivator and wind breaks   *Bottom Left:* 12-row cultivator
Right: Cotton picker used in off season as a cultivator in South Carolina

CROP ROTATION

Left to Right: Crop rotation   In June cotton follows wheat in Southwest Georgia

Left to Right: Soybeans planted in early June follow wheat in a four and four skip-row pattern with cotton
No tillage and sprinkler irrigation work well for Bill Smith of Cameron, South Carolina

Left to Right: Two rows of soybeans planted in four-row skips with cotton   No-tillage crop rotation

Left to Right: A cotton/corn rotation in North Alabama   A cotton/alfalfa rotation in Arizona

FERTILIZATION

Top Left to Right: In the forties fertilization was "back breaking" work   Lime spread in fallowed four-rows of skip-row
Bottom: Applying liquid nitrogen

Left to Right: A tractor with a fertilizer rig heads back to the headquarters at sunset
Application of anhydrous ammonia, the cheapest form of nitrogen

Top Left to Right: Chicken litter stock piled for a Missouri cotton grower   Farmers and "city folk" breakfast at The Idle Hour Restaurant in Scotland Neck, North Carolina   *Bottom:* One of the largest liquid fertilizer rigs

Top Left to Right: Mississippi Chemical Corporation in 1962   Jerry Clower, humorist and one-time Mississippi Chemical salesman
Bottom: An insecticide is applied by a modern agricultural plane

HOES & HERBICIDES

Top Left to Right: Hand choppers in "day crop" "block out" seedlings   An old school bus transports choppers to the country
Bottom Left to Right: A single chopper "rogues" the field of weed pests note the "croaker sack" for taking weed pests from the field
"Brushed through" by hand choppers

Top Left to Right: In the early fifties, a comparison of rows banded with pre-emergence herbicide, with one that was not "Wow! What these pre-emergence herbicides can do"   *Bottom:* Healthy seedlings, clean drills

Top Left to Right: In 1965, Reno Borgognoni mixes Karmex for "lay-by" application
Slide type applicator for "lay-by" herbicides   *Bottom:* The "new look" in cotton choppers

Top Left to Right: Men spray an over-the-top herbicide   Tracy Reed applies herbicide spray on Texas farm
Bottom: Misti Henley sprays Johnson grass in Tennessee Valley

Top Left to Right: Weed control by flame cultivation   Wick bar used to rub Roundup on Johnson grass
Bottom: Researchers at the Delta Branch Experiment Station, Stoneville direct a post-emergence herbicide in foam

Top Left: Ground application of Roundup to Roundup Ready cotton   *Bottom Left:* Aerial application of Roundup
Right: On May 12, 2001 with perfect stands and weed control the Logans of Louisiana say,"let's go deep sea fishing on the Coast"

IRRIGATION

Left to Right: THE Delta Queen on the Mississippi River, the source of great aquafiers across the Delta country
A pretty "county fair" stalk of cotton grown in Alabama

Irrigation in the "good ole days"
Top Left to Right: Center-pivot irrigation system   Solid-set irrigation system   *Bottom:* Side-wheel roller sprinkler

Top Left to Right: Irrigation well   Furrow irrigation system with siphon tubes out of a dirt ditch, Greenwood, Mississippi
Bottom Left to Right: A furrow irrigation with siphon tubes out of a concrete ditch, Arizona   Furrow irrigation system with poly pipe

Top Left to Right: Dry land cotton in Texas   Subsurface drip irrigation in Texas
Bottom Left to Right: Furrow (left) versus drip irrigation   Arizonian, Howard Wuertz, pioneer of subsurface drip irrigation

Left to Right: Extension Service farm tours   Linear-move irrigation system

Top Left to Right: South Carolinian, Ed Rast and wife, observe their irrigation system
The world's largest center-pivot irrigation system that covers a thousand acres in one revolution
Bottom: This center-pivot system is built to clear oil wells

Top Left to Right: Cornering arm on a center-pivot system   Circular rows are marked off with a center-pivot system before planting
Bottom: Drop water spray nozzles for irrigation conserve water

SKIP ROW

Top Left to Right: Two rows of soybeans planted after a good growth of cotton   Two and one skip-row cotton in red dirt of Alabama
Bottom: Eight-row cultivator of the sixties

Top Left: Ultra-narrow and normal cotton on the Charles Coghlan Farm at Scott, Mississippi
Bottom Left: The harvest of ultra-narrow row cotton in the seventies   *Right:* "Cotton picking combines" at Stoneville, Mississippi

Top Left to Right: Sencorp Stripper Header on the front of a tractor blows seed cotton into a Mule Boy   A Stakhand "calved" a seven-bale module of seed cotton   *Bottom Left to Right:* Cotton stripper   Ultra-narrow-row cotton in Texas

Left to Right: Dr. Steve Husman, of Arizona, conducts tests comparing ultra-narrow-row cotton and conventional row cotton
Stalks of old ultra-narrow cotton row stalks

Left: Four and one skip-row cotton in Arizona
Top and Bottom Right: One-row configurations in two and one skip-row fields in the Mississippi Delta

Top Left: Outside rows produce more cotton than interior rows   *Bottom Left:* In the seventies, Betty Pendergrass Scott in 100% cotton
Right: In 2001 Brent Shaw, North Alabama, tried ultra-narrow row cotton in a diamond shape

STAGES OF COTTON

Seedling cotton

White bloom

Red bloom

Green bolls　　　　　　　　　　　Cracked boll　　　　　　　　　　　Open boll

INSECT CONTROL

Opposite page, Jack Shannon, "duster pilot" takes off in his Stearman
Top Left: Snookie Spaggiari prepares the insecticide "Sweet Kill" for ground sprayer   *Bottom Left:* Growers applied "Hot Sauce" to give the worms stomach ulcers   *Right:* Calcium arsenate was dusted on this field for boll weevil control

Top Left to Right: Ag pilot applies insecticide   A wide-bodied ground sprayer for skip-row cotton
Bottom: Home-made ground sprayer

Boll weevil

Cotton bollworm

Top Left to Right: Harris Barnes, III inspects plants for insects   Bill Pellum is motorized for skip-row cotton
Bottom Left to Right: In 1957 Dr. Ivan Myles and Jack Oakman show how insecticides help "stick" fruit
Grower, Charles Youngker of Buckeye, Arizona, goes over his insect inventory with his consultant Al Linguel

Left to Right: Controlled droplet applicators   Insecticides applied from a cultivator
Previous page, "Cotton" Carnahan, veteran agricultural pilot of Clarksdale "starts his engine"

Left to Right: Modern "needle-nosed" aircraft   Helicopter applying insecticide in "tight" spots

MATURE COTTON

Top Left to Right: 1964 urban sprawl takes over good farm land   The "Church Cut", South Mississippi Delta   *Bottom Left to Right:* North Carolina cotton takes over some tobacco land   In the past family cemeteries were found in the middle of some cotton fields

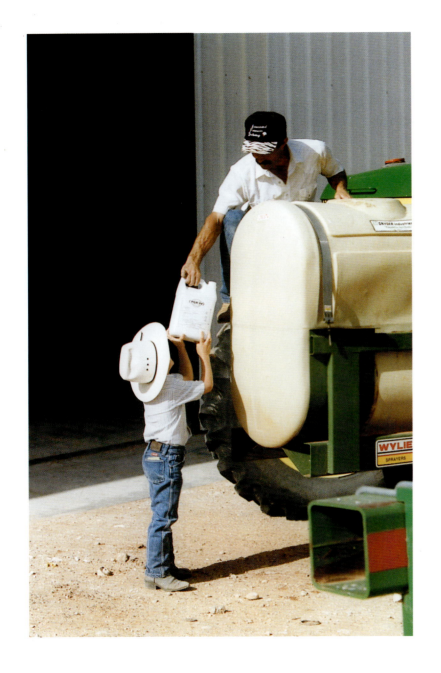

Left: Future farmer, Dustin Nelson helps his dad, Larry   *Top Right:* Mississippi State Extension sponsored field tour
Bottom Right: Larkin Martin, Managing Partner of a large North Alabama farm

Left to Right: In North Alabama, marked cotton rows help machine operators   The John Howard Farm, Deep Run, North Carolina

Previous page, The Harwell Home, The Columns, Florence, South Carolina
Left to Right: A uniform stand   Cotton near Yuma, Arizona

OPEN COTTON FIELDS

Opposite page, Famous Deer Creek originates in Scott, Mississippi
Left to Right: Outside row, two and one skip-row cotton   The west side of San Joaquin Valley, California

Left: Deer Creek in the dry summer of 2000
Right: Georgia extension leader inspects cotton crop of Percy Dixon, Burke County grower

Left to Right: Jim Hanson and sons, Philip and Erik, stand in three-bale plus cotton near Corcoran, California   Sunset on a great top crop of cotton

Top Left to Right: Stan Andrew, Gila River Farms, Sacaton, Arizona, in a field of four and one cotton   A "Cabin in the Cotton", Yazoo City, Mississippi   *Bottom:* Mark Borba, Borba Farms, Riverdale, California in a field of Chembred CB7 Hybrid

Top Left to Right: Larry Fenster, Delta Airlines Pilot and part-time cotton grower   Terry white, Altus, Oklahoma
Bottom Left to Right: Ann Ruscoe, first female County Extension Agent in Mississippi   Miss Southern 500, Cindy McDowell Holt, Greenville, South Carolina, in 100 % cotton   *Next page,* The beauty of the cotton harvest in West Tennessee

DEFOLIATION

Left to Right: A crop duster applies "Black Annie" one of the best defoliants ever made   An old Stearman in the sixties clears old plantation home

Left to Right: Liquid defoliant applied with high clearance ground sprayer
The photographer follows instructions - Don't shoot until you see the whites of their eyes

Previous page, Defoliation at sunset
Left to Right: A Mid-South Farm Manager, Johnny Lawrence supervises all operations   Liquid defoliant applied to cotton field

HAND & MECHANICAL PICKING

Top Left and Right: Hand picking always made for the best grades   *Bottom Left to Right:* The James Johnson Family, Baugh Plantation, Sherard, Mississippi, picks their own crop   In the fifties, before herbicides sometimes more grass and weeds than cotton

Left to Right: "Weighing up" and "shaking out" in the forties   Loading out seed cotton from a tenant shed to haul to gin

Top Left to Right: First mechanical picker, International Harvester one-row picker   Mechanization of the cotton harvest took over in the late forties and early fifties   *Bottom:* Two one-row pickers with some seed cotton were abandoned for the new two-row versions

Top Left: New John Deere two-row picker in good cotton   *Bottom Left:* High and low drum, two-row Allis-Chalmers pickers in the fifties
Right: Four and five bale/acre cotton on Gila River Farms, Sacaton, Arizona   *Opposite page,* "Cotton kids"

Top Left to Right: Rob Lewis, Coahoma County Mississippi Extension Agent with Leon Bramlett, grower
Model, Nita Abraham Ross, examines good machine picked cotton   *Bottom:* At sunset, seed cotton unloaded rather than dumped

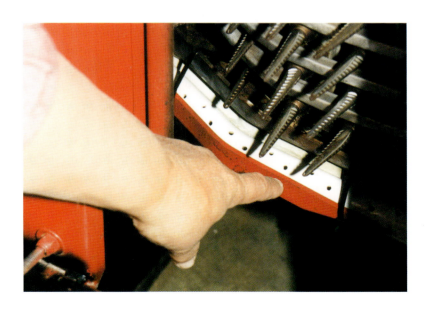

Helps in the harvest
Top Left to Right: A little push in wet spots   Dual and rear wheel assists   *Bottom:* Air to keep locks floating upward

Left to Right: Mike Moore of Waddell, Arizona, a one time producer of brown and green seed cotton
"Fast colors" in fabrics woven by Sally Fox

Top Left to Right: Pima cotton, Harquahala Valley, Arizona   Great cotton in South Delta Mississippi
Bottom: Irrigated cotton field on the South Plains of Texas that produced 1200 pounds of lint per acre after five hailstorms

Left to Right: Harvesting bottom defoliated cotton in Arizona   The last "rows" of fall on Mitchener Farms, Sumner, Mississippi

Top Left to Right: The world's widest picker   Seven John Deere six-row pickers work on the Flowers Farm, Coahoma County, Mississippi
Bottom: In 2001, Case International introduced their six-row picker

Left to Right: Tasha Wells, GPS specialist works with Ernie Iler, Farm Manager, Pineland Plantation, Southwest Georgia on yield monitoring with a mechanical picker   Lori Franklin Johnson, Hood Farms, Perthshire, Mississippi, actually runs a 10-row 30-inch row planter and this five-row picker *Opposite page,* New six-row pickers in two bale plus cotton as on the David Wildy Farms near Manila, Arkansas

HUMAN INTEREST

Left: Happiness is Prince Albert in a pipe
Top and Bottom Right: In the fifties and sixties, Hays Brothers "cleared" houses from thousands of Mid-South acres

Top Left: A new six-row cultivator and new homes for the machine operators   *Bottom Left:* A new home and no wood to "tote"
Right: Nothing finer than Good Gulf Gas and Bull Durham tobacco

MODULES

Excellent cotton - crowded gin yards lead to dumping on the ground

Seed cotton was blown into barns and had to be sucked out when the gins caught up

Left to Right: In the sixties, prior to the development of the module builder, cotton growers, especially in Mississippi and Texas where using wood and steel Cotton Caddies to shape ricks of seed cotton, sometimes a mile long, on the turn roads.

Left to Right: A great day, came about in 1972, when the module builder and transporter were demonstrated to growers in the Rio Grande Valley. Most say the module builder was third in importance, just behind the gin and the mechanical picker in the harvest of cotton

Left to Right: Arkansas cotton grower and ginner, Larry McClendon, was able to do a better job securing the covers on his modules by laying twine across the center as the module was being made   The "school solution"—-a neat, well packed module of seed cotton
Previous page, The end of the day

Left to Right: To avoid hauling modules at night this Louisiana gin "calved" them on to 17 flat rail cars sufficient to last the night
These modules parked closely along a California irrigation ditch indicate either a high-yielding crop, long rows or both

Left to Right: The sunset, the modules and the mountains all add to the beauty of agriculture in the Harquahala Valley of Arizona
"Calving" modules at sunset on a Mississippi gin yard

Left to Right: This Mississippi grower posts his prayer on one side of his modules located just off Highway 6 near Marks
"Golden Retriever" retrieves seed cotton left under modules

GINS

Left to Right: In the fifties a one-bale trailer on the way to a gin   An old conventional "suck pipe" gin

Left to Right: Trailers of seed cotton lined up for ginning
A new "Disney World Colors" multicolored type gin on King and Anderson, Inc., Clarksdale, Mississippi

Previous Page: In 2001 1,142 modules at one time on the yard Bobo Moseley gin, Lyon, Mississippi
Right: Cottonseed, bales, modules and mountains as seen from the yard of a Harquahala Valley, Arizona gin

Left to Right: New roller gin under construction in Arizona   New high capacity gin in the Mid-South
Opposite page, The world's largest 10-stand gin in Arkansas

BALES

The Trail of the Ugly Bale

Sampled at the gin and the compress   Compressed   Transported to the textile mill or the ship for export

Left to Rght: This trail came to an abrupt end with the universal density bale   Neat bales of Pima cotton await shipment from the gin yard of the J.G. Boswell Company, Corcoran, California

IN THE WINTERTIME

Previous page, A no-till start for the 2002 cotton crop
Left to Rght: Snow makes for mellow soil and excellent stands   Cotton fields surround this Delta lake, full of fish and cypress trees

Left to Rght: What do Mid-South cotton growers do in the wintertime? Attend auctions, hunt deer, ducks and rabbits
Next page, Mallards paired

PEOPLE

Top Left to Rght: Chauncey Taylor, cotton grower, inventor and manufacturer of "Little Bales"
A cotton tour group from People's Republic of China looking over an Arizona field
Bottom: James Lee Adams, one of the first to use computers and alligators on his Camilla, Georgia farm

Top Left to Rght: Two outstanding cotton merchants of the Cotton Belt over the past 50 years
Bill Connell of Connell Brothers, Clarksdale, Mississippi   Billy Dunavant of Dunavant Enterprises, Memphis, Tennessee
Bottom: The 1985 Maid of Cotton, Michelle Pitcher from the cotton producing state of Missouri, "pulls cotton" in the classing room of the Joseph Walker & Company in Columbia, South Carolina

Thanks to our entire family, four children and ten grandchildren as pictured at Christmas 1998.

And to my immediate family (left to right) Jamye, Dudley, Jim, Harris III, Jamye, Jr., and Harris. This picture taken by world-renowned ag photographer, John McKinney of *Progressive Farmer* showed how our family boosted cotton by wearing all cotton.

ILLUSTRATION DATA

LAND PREPARATION

PAGE 14, Just after World War II much of the cotton land was prepared in the Mid-South by men and mules such as this four up, pulling a two row disk harrow.

PAGE 15, Top Left: Beautiful Delta land is being turned here with a small tractor and a two-bottom turning plow. Top Right: On a large plantation, several tractors pulling two-row field cultivators, prepared the land to be set up in beds by three-row middle breakers. Bottom Left: A North Carolina grower breaks his high-organic soil with a four-bottom moldboard plow, followed by sea gulls looking for worms. Bottom Right: This operator sets up four beds at a pass with a disk hipper.

PAGE 16, Left: In the Mid-South, before the planting, the rows were drug down to the desired height with a four-row Do-All, composed of a flip-flop stalk cutter, a section harrow and a drag board. Top Right: In the fifties, Cliff Porterfield, chemical salesman and consultant, calibrates (a new word for the cotton farmer) one of the first broadcast spray applicators. Bottom Right: A herbicide is broadcast with this ground sprayer. The two large tanks were for the herbicide, and the small, front-mounted tank was for an insecticide, a fungicide, fertilizer or a plant growth regulator.

PAGE 17, Left Top: With the advent of wide-row equipment, it was common in the nineties to see 12-row hippers or busters setting up 12 rows at a pass. Left Bottom: The modern way of running water furrows in the fields. Right: In the springtime, the cotton grower was tempted to hunt turkey instead of preparing land or planting cotton.

PAGE 18, Left: A South Plains Texas grower breaks his sandy soil with a six-row "switch bottom" plow. Right: On the South Plains of Texas, several rows of wheat are planted in the middles and sides of beds to lessen the loss of soil during the winter and to protect the young cotton seedlings as they emerge in the spring.

PAGE 19, Left: The Pioneer of the stale seedbed, Ray Young of Louisiana, does all of his land preparation after the cotton harvest by cutting stalks and hipping up the old beds. After killing the winter vegetation in the spring he plants directly into the beds. Right: Land preparation at planting time usually went past sunset and "far into the night".

PLANTING

PAGE 20, Top Right: In the late forties came the first four-row, "running board" planters, with levers to set depth of seed for each row. Tanks, with herbicides or insecticides are being used. Bottom Left: The first four-row planter on Baugh Plantation, Sherard, Mississippi, consisted of old 55-gallon drums on each side for the herbicides and water. The old truck featured a 1000-gallon water tank. Bottom Right: In the fifties, CIPC, this one HO-NO-MO, was one of the most popular pre-emergence herbicides. The Deltapine variety cottonseed were packaged in good looking, 100 % cotton bags which were coveted by all for bed spreads, curtains and many other uses.

PAGE 21, Left: One of the first John Deere riding eight-row planters. The levers are for adjusting the seed depth on each row. Right: This International Harvester four-row was a riding planter and perhaps one of the most "flimsy" ever built. Thurston Pellum, Farm Manager, Oakhurst Company, checks the depth of the seed as the operators anxiously await his call.

PAGE 22, Left: The mule and hay barns of the "good ole days" on Stovall Farms, Stovall, Mississippi, overlook the planting of cotton adjacent to the headquarters. Right: Most of the cotton planted along the Mississippi River is planted on the "land side" of the levees. However, in some areas in Tennessee, Arkansas, Mississippi and Louisiana there are cotton growers who take chances each year that the river "will not rise" after the seed are dropped into the soil.

PAGE 23, Top Left: A four-row planter, dropping seed in a skip-row fashion of two and two is seen through the "legs" of a Farm Manager. Bottom Left: Three indications of the time to plant cotton – when the pecan trees begin to put out leaves – when the ladies, who are fishing, sit on the soil rather than on the coke boxes, and when the soil temps read right. Donna Moore Surholt (left) and Mary Margaret Humber Polles wearing all-cotton, take temperatures of soil to determine if all is right for the germination of the seed. Right: Planting time at sunset on Moon Lake created when the Corps of Engineers took out most of the curves in the Mississippi River.

PAGE 24, Left: Cottonseed being dropped into a stale seed bed in Louisiana. Right: Cotton being planted in the Bootheel of Missouri.

Wheat is planted in the middles after the harvest to save the soil during the winter and spring. Also it serves as a windbreak for the seedling cotton after emergence.

PAGE 25, Four of the latest John Deere tractors with 12-row planters drop cottonseed on the Flowers Farm in Coahoma County, Mississippi.

PAGE 26, Top Left: The EPA would like to see all cotton growers wear this clothing at planting time. Top Right: A large modern planting operation with a truck full of chemicals, tanks of liquid fertilizer and water and a 12-row planter with solid material applicators. Bottom: Tim Patrick, who farms in West Tennessee near Covington, plants cotton in the no-tillage style in the rolling hills.

PAGE 27, Top Left: A few growers punch their empty paper bags over a spindle as they fill the planters and burn the residue as they depart. This prevents the paper bags from being scattered about the fields to interfere with cultivation and other practices during the growing year. Top Right: Jimmy Hargett of Bells, Tennessee manufactured the widest cotton planter and picker in the world. The rows of cotton are spaced 30 inches with 60-inch skips. He also made the wide row, six-row, skip-row picker used by him and Hood Harris of North Alabama. Bottom: What a planting season is all about – good men and women, machinery, seed, chemicals, fertilizers and weather.

SEEDLING COTTON

PAGE 28, Half the battle is obtaining a good healthy stand of cotton seedlings.

PAGE 29, Left: The Mississippi Delta – cypress trees and a perfect stand of hill-dropped cotton. Right: A "cabin" in the Tennessee Valley of North Alabama.

PAGE 30, Left: The "old timers" have always contended that a cotton farmer can make more cotton on a curved or crooked row rather than a straight one. Green uniform cotton adds to the beauty of the red dirt of the Tennessee Valley of North Alabama. Right: A cotton field of skip-row four and one, with great vitality and uniformity.

PAGE 31, Top Left: Beth Bobo, Clarksdale, Mississippi, shows what cotton plants should look like in early June when protected by wind breaks such as woods or grown up ditches. Top Right: Greater joy hath no grower than to walk through fields with perfect stands of healthy, weed-free cotton. Bottom: Magic! No-till cotton in North Alabama – Burn down by Roundup.

PAGE 32, Left: Not the Sahara Desert but the Bootheel of Missouri, where fields of sand and seedling cotton are protected by green or "killed wheat". Right: Great protection. Some Missouri growers prefer to let the wheat grow until prior to cotton emergence at which time it is killed by one of the "burn down" herbicides.

PAGE 33, Left: Missouri grower, Ron Lea likes the way the wheat has saved both his soil and his cotton seedlings. Right: In the Bootheel of Missouri these cotton seedlings are protected by "killed wheat".

CULTIVATION

PAGE 34, Left: A four-up pulls a two-row riding cultivator in the late forties on Kline Planting Company, Alligator, Mississippi. Right: In the late forties, Anthony Harris cultivates his crop on Baugh Plantation, Sherard, Mississippi with a double shovel.

PAGE 35, Left: In the sixties five tractors with four-row cultivators did about the same work as one large tractor with 12-row cultivator would today. Note the tanks for the water and herbicides and the "air conditioning by umbrella". Right: A four-row cultivator with slides protects the treated band for the application of post-directed herbicides on the Elkhorn Plantation of King and Anderson, Inc., Clarksdale, Mississippi. The tall building in the background served as a gin and later as a barn for mules and hay. "Old timers" say that the owner of this farm would sometimes go into the cupola with binoculars to check on the work that was going on in nearby fields.

PAGE 36, Left: A PTO type cultivator pulverized the soil and left a smooth band for the post-directing of herbicides. The row marker, if watched closely, prevented the chewing up of the cotton plants. Right: A cotton grower supervises the cultivation of his cotton crop to make sure the "treated bands of herbicides" are preserved.

PAGE 37, Left: This was one of the most popular four-row cultivators of the sixties. Three wide sweeps on the rear made for a double cultivation and a wiping out of the prints of the tires. Right: The operator directs a herbicide into the drill with Bell-type applicators. "Sweeps" take out weed pests in the middles.

PAGE 38, Left: A service truck of the mid-sixties was an old tractor on which was mounted a toolbox with tools and a water keg. Most of the tractors in those days were fueled with LP gas. This one is hitched to a 1000-gallon tank with propane gas. Right: Before the advent of strong, long tractor axles growers were forced to cultivate their skip-row two and one with one-wheel between the rows and one wheel in the skip.

PAGE 39, Top Left: Cultivation taking place on the KBH Experimental Farm in the early seventies with a ten-row cultivator. There is a four-row windbreak for every twenty rows. Bottom Left: A modern twelve-row cultivator. Right: This South Carolina grower took advantage of a cotton picker and its power unit by using it in the off-season to cultivate six-rows of cotton. The unit with its music, air conditioning and comfortable seat was very popular with his operators.

CROP ROTATION

PAGE 40, Left: A crop rotation of peanuts to wheat to cotton is very evident here. Right: In June, Ernie Iler, General Farm Manager, Pineland Plantation, Newton, Georgia, examines this uniform, healthy field of cotton planted behind a wheat harvest.

PAGE 41, Left: In the past a few Mid-South growers planted two to four rows of soybeans in the four-row skips after the cotton plants had made a good start. Right: In South Carolina, Bill Smith burned down a cover crop and planted cotton in the no-tillage fashion. And it all works better with center-pivot sprinkler irrigation.

PAGE 42, Left: Two rows of soybeans planted late in four-row skips. Right: In no-tillage it is easy to see the crop rotation that has taken place.

PAGE 43, Left: Nothing can be finer than this cotton/corn rotation in the Tennessee Valley of North Alabama. Right: In Arizona a crop notation of cotton and alfalfa is very popular.

FERTILIZATION

PAGE 44, Top Left: Solid fertilizer in the forties meant a lot of hard backbreaking work. When delivered, prior to the growing season, the heavy bags had to be loaded into a shed or old barn. Many times the weight of the fertilizer caused the floor joists to break. The fertilizer "sweated" so much that the bags became hard as cement. Each bag had to be dumped into the fertilizer distributors, some "stout" enough to distribute the hard material. Top Right: When a cotton crop, planted in the four and four skip-row-fashion was harvested, lime, if needed, was spread on the fallowed part to be planted to cotton the following year. Bottom: It was a great day for the cotton grower when liquid nitrogen and mixed goods came on the market which did away with a lot of the hard work associated with fertilization of the cotton crop.

PAGE 45, Left: A fertilizer rig is pulled back to the headquarters at sunset-quitting time. Right: Anhydrous ammonia has been, and will continue to be, one of the cheapest forms of nitrogen for the cotton grower. It was developed for ag use by the late Dr. W.B. Andrews of Mississippi State College (now known as Mississippi State University).

PAGE 46, Top Left: Near to the poultry industry are the cotton growers in the Bootheel of Missouri. After the crop is planted they begin to stock the chicken litter on turn roads and other spots for the following crop year. Consultant Barry Aycock stands on top of some of the stocked fertilizer materials. Top Right: Farmers breakfast with "city" friends at the Idle Hour Restaurant, Scotland Neck, North Carolina. Bottom: One of the largest and most modern liquid fertilizer rigs, does a good job of injection on a farm near Greenwood, Mississippi, known as the World's Largest Inland Cotton Market.

PAGE 47, Top Left: In 1948, Mississippi Chemical Corporation, Yazoo City, Mississippi, was first organized as a cooperative and owned by 600 farm families. The first anhydrous ammonia at Mississippi Chemical was produced in 1951. Dr. W. B. Andrews, Mississippi State College (now Mississippi State University), was responsible for the technology that made it possible for the farmer to inject this most economical nitrogen material into the soil. Prilled ammonium nitrate, that flowed like marbles from a paper sack, helped farmers more effectively utilize this fertilizer on their crops. (This photo was taken in 1962) Top Right: Jerry Clower of Yazoo City, Mississippi, was first a seed corn salesman before joining Mississippi Chemical Cooperation where he sold many tons of fertilizer as he told his clean, funny stories all over the USA. Bottom: In 1992, a modern ag aircraft applies liquid potash on a cotton field on Mitchener Farms, Sumner, Mississippi.

HOES & HERBICIDES

PAGE 48, Top Left: In the early fifties, hand choppers were trucked in from nearby towns just to "block out" perfect stands of drilled cotton. Top Right: A familiar sight in the forties and fifties was an old school bus or truck parked on the end of the rows where water, extra hoes, lunches, files, etc., were kept. For the most part this day labor put in a 10-hour day, starting at 0600, taking an hour for lunch which was brought in paper bags or gallon tins, and leaving at 1700. Bottom Left: Many times there were not enough weed pests in a given field to warrant a "gang" of choppers. In this case a man was paid to "rogue" the fields, to go from spot to spot to chop mature weed pests. A "croaker sack" was used to carry out weed pests that had made or were about to make seed. Bottom Right: Obviously a group of hand choppers have "brushed through" this field taking care of the small weed pests in the drill.

PAGE 49, Top Left: 1952 was the first year that pre-emergence herbicides were used in cotton fields. This is what an untreated portion of a row looked like, as compared to treated rows on either side. Top Right: The "school solution" for pre-emergence herbicides – healthy seedlings and clean drills. Bottom: In the early fifties came the first pre-emergence herbicides for cotton. Growers could not believe their eyes when they saw that a chemical would take out most of the grass and weeds, and leave healthy cotton seedlings.

PAGE 50, Top Left: In 1965, Reno Borgognoni mixed Karmex and water as a lay-by solution to spray on his cotton middles. Top Right: Prior to the cotton plants "lapping" the middles, many farmers made their last pass through their fields applying a so-called, lay-by herbicide from a slide-type applicator. A single flood type nozzle per middle distributed the herbicide across the middle and to the base of the cotton plants on each side. Bottom: For years now, Heaton Farms of Clarksdale, Mississippi, has hired high school boys to "chop" cotton from 0600 to 1200 five days a week when there were weed pests in the fields. The group is usually commanded by one of the high school coaches. During their time together there is quite a bit of "fraternity" developed. The summer ends with watermelon at the boss's pool! The mode of transportation has greatly improved over the past 50 years!

PAGE 51, Top Left: In 1969, one of the best ways to control Johnson grass was with men seated on the front of a tractor, with a good herbicide and a spray gun. Top Right: In a pre-harvest spray, Tracy Reed on Jeff Shaddon Farm, Abernathy, Texas, sprays the 1999 weed pests in her crop to prevent contamination of the lint and the making of seed by the weed pests. Bottom: To make some spending money, Misti Henly, sprays Johnson grass from her four-wheeler.

PAGE 52, Top Left: Prior to the cotton grower using herbicides for post-directed weed control, many turned to flame cultivation. Using propane gas through flame cultivators a flame was shot across the drill at the proper speed about once a week with good control of the weed pests and very little damage to the cotton plants. Flame cultivators were first tested at the Delta Branch Experiment Station in Stoneville, Mississippi, in 1943. Butane and propane were first used as fuel for the flame in 1945. Top Right: Johnson grass has been, over the years, one of the most important weed pests for Southern farmers. Wick bars that rubbed a herbicide like Roundup on the tops of the grass heads, have been very successful. Bottom: In 1968 researchers at the Delta Branch Experiment Station, Stoneville, Mississippi, were experimenting with the delivery of post-emergence herbicides to the drills of the cotton with a colorful foam.

PAGE 53, Top Left: Ground application of Roundup to Roundup Ready cotton in 2001. Bottom Left: Aerial application of Roundup to Roundup Ready cotton. Right: It used to be in the "good ole' days Danny Logan and his son, Stephen of Gilliam, Louisiana, would be fighting insects, weeds and bad stands early on. However, on May 12, 2001, because of good fertilizers, herbicides, insecticides, fungicides and Bt cottons resistant to over-the-top applications of herbicides, the father/son team was preparing to go deep sea fishing in the Gulf of Mexico.

IRRIGATION

PAGE 54, Left: The Delta Queen on the Mississippi River, the source of great aquafiers in the Mid-South. Right: Tiger Bridgeforth of Darden Bridgeforth and Sons Farm, Tanner, Alabama admires a pretty "county fair" stalk of cotton that has been the recipient of good weather, irrigation, insect and weed control. Note the fruit set on out to the fifth and sixth position on some limbs.

PAGE 55, Top Left: One of the first center-pivot irrigation systems to be used was this water-driven Valley equipped with high-pressure impact sprinklers. These systems were last manufactured and sold in the late seventies. Top Right: Solid-set irrigation systems were responsible for some of the first irrigation of cotton in the Delta Country. It was made up of aluminum pipe, risers and high-pressure impact sprinklers. Since these systems had to be "hand transported" under wet conditions from place to place during the hot summer time, old timers claim to have one or more of these systems was the best way in the world to run good labor from the farm. Bottom: Side-wheel roller sprinkler systems were very popular in the Western areas. Equipped with high-pressure impact sprinklers the system was moved every hour or so to a new location by a gasoline engine powered drive, coupled to a chain drive and sprockets.

PAGE 56, Top Left: In 1957 this well, capable of delivering 1200 to 1500 gallons per minute was used on the Lea Plantation near Dublin, Mississippi. A reservoir enabled the water to be sent in three directions. For the most part flumes were constructed on pads with the highest elevation so that the water, lifted out by siphon tubes, could be sent down the middles on both sides of the ditch. Top Right: One of the first furrow irrigation systems was done on the Hardeman Plantation, near Greenwood, Mississippi. Fields were formed to give a .1 to .2 grade. The 1.5-inch siphon tubes after starting, were placed one to the middle. Bottom Left: Furrow irrigation with concrete ditches and siphon tubes is done on Gila River Farms, near Sacaton, Arizona. Bottom Right: Where land has been formed with a slight grade, furrow irrigation with poly pipe has become very popular in the Mid-South and Southeast.

PAGE 57, Top Left: Dry land cotton on the South Plains of Texas. Jim Risley produces about a half a bale of cotton per acre on dry land, compared to 1.5 bales per acre on irrigated cotton. Top Right: Subsurface drip irrigation makes for large open bolls and higher yields per acre on the South Plains of Texas. Bottom Left: There is a big payoff in subsurface drip irrigation on the South Plains of Texas. Many growers claim drip irrigation permits their "bad" water to go twice as far and at the same time double their cotton yields. Note the drip-irrigation cotton on the right, as compared with furrow-watered on the left about a-bale per acre difference. Bottom Right: Howard Wuertz, Coolidge, Arizona, is a pioneer in subsurface drip irrigation. Here he shows off cotton where drip lines are placed in every other middle rather than under each row.

PAGE 58, Left: Farm tours, sponsored by researchers and extension personnel of land grant universities, concentrate on irrigation, varieties, insect and weed control and other production practices. Right: This side roller is in reality a linear-move irrigation system that takes care of rectangular fields very nicely.

PAGE 59, Top Left: Ed Rast and his wife, Jeanne, of Cameron, South Carolina, enjoy the sight of their center-pivot irrigation system from the porch of their home. Top Right: One of the world's largest center-pivot irrigation system is a Valley located on the Owen Brothers Farm near Tunica, Mississippi. It irrigates better than a 1000 acres of cotton in a revolution that takes five days to make the 4.5 miles. It is three-fourths of a mile long and is made up of 26 towers. Two 16-inch wells manifolded together, furnish the needed 3,800 gallons per minute. There are not many systems that large to be found in cotton country for not many areas have aquafiers capable of supplying that much water. Bottom: On the South Plains of Texas "High Rise" center-pivot systems clear oil wells that "clutter" the cotton fields.

PAGE 60, Top Left: A cornering arm on a center-pivot system can add as much as 285 feet to the coverage. Top Right: Many cotton producers, on the South Plains of Texas, will mark off their rows with the wheels on a center-pivot system before planting. This enables them to plant in circles and be able to drag water nozzles down each middle. Bottom: Drop water spray nozzles irrigate the cotton crop of this Texas grower as he inspects some plants for insects.

SKIP ROW

PAGE 61, Top Left: Rather than fallow the strips all summer, a few growers elected to plant two rows of soybeans after the cotton plants had made good initial growth. Top Right: In the Tennessee Valley in North Alabama water for irrigation is hard to get because of the underground rock formations. Many have adopted the two and one skip-row configuration with great success. With average conditions many feel that yields per ground acre will be as good as those from solid-planted fields. Bottom: In 1969 this eight-row cultivates four rows of cotton and four rows of skips in this two and two configuration. Fenders are used to prevent dirt being thrown up on the treated band and as a mount for directing herbicides.

PAGE 62, Top Left: On the Charles Coghlan Farm at Scott, Mississippi, yield tests were run with conventional row spacing versus those of ultra narrow row cotton. Right: 1964 saw the first use of ultra narrow cotton at the Delta Branch Experiment Station, Stoneville, Mississippi. Dr. Gordon Tupper is pictured operating the Ben Pearson "cotton picking combine". Some of the cotton was planted in 6-inch rows and some seed broadcast from a plane. There were eight of the Ben Pearson "cotton picking combines" manufactured and they were used later as a machine for second picking and stalk cutting. Bottom Left: Cottonseed was broadcast-flown with a plane on some farms.

PAGE 63, Top Left: On Hood Farms, Perthshire, Mississippi, they attached a Sencorp Stripper Header to the front of a tractor. The seed cotton was blown into a large gin type cleaner and extractor, then into a large KBH Muleboy, which has the capability of dumping into a conventional module builder. Top Right: The Hoods also used a similar rig, however the seed cotton was blown into a Stakhand, originally designed to make modules of hay. In this case the Stakhand "calved" seven-bale modules of seed cotton. They were able to put two of these mini-modules together for hauling, ginning, etc. Bottom Left: Gin brushes bolted to the reels of stripper headers have greatly increased the efficiency of the harvest of ultra narrow row cotton. This cotton stripper was one of the harvesters used on Hood Farms. Bottom Right: Some of the best ultra-narrow-row cotton seen was on the Brad Martin Farm near Edmonson, Texas. The reel had gin brushes attached which helped push the seed cotton from the fingers into the auger. The excellent job done by the stripper can be seen.

PAGE 64, Left: In the Casa Grande, Arizona area, Dr. Steve Husman conducted tests comparing ultra-narrow-row cotton to conventional-row spacings. He found over a two-year period that the ultra-narrow-row cotton system resulted in a 77-pound lint yield increase and a $69 an acre lower production cost. There was also a trend towards lower micronaire in the ultra-narrow-row cotton system. Because of the issues and challenges in the ultra-narrow-row cotton production Husman is concentrating more of his efforts now on the production of cotton in twin rows. Right: For a number of reasons, Mike Murphy, the Tennessee Valley of North Alabama, "turned his back" on ultra-narrow-row cotton for the most part and has embraced conventional row widths again. Note the old ultra-narrow-row cotton stalks, in the middles of his conventional cotton in 2000.

PAGE 65, Left: Four and one, skip-row cotton on Gila River Farms near Sacaton, Arizona. They like this skip-row pattern for yields from outside rows sometimes double that of the adjacent interior rows. This pattern also serves as a good "marker" for cultivators, spray machines and two or four-row cotton picker operators. Right Top and Bottom: Bruton Farms of Hollandale, Mississippi, has 5,000 acres of cotton and has shifted from five four-row cotton pickers to three, six-row pickers. Yields range from 850-1,150 pounds lint per ground acre. They irrigate with furrow irrigation which gives them another 300 pounds of lint over dry land cotton. During acreage restrictions, they were able to use more of their good sandy cotton land by using one single row in their skip-row pattern.

PAGE 66 Top Left: Most years "outside rows" make twice as much cotton as the interior rows. Bottom Left: Betty Pendergrass Scott shows that girls in 100% cotton dresses were just as pretty in the seventies as now. Right: Brent Shaw, who farms in the Tennessee Valley of North Alabama, experimented with UNRC in 2001 by planting sets of 38 inch rows at 30 degree angles. He was well pleased with the almost two-bale yields but, will not continue the practice, due to stripping and ginning problems.

STAGES OF COTTON

PAGE 68 - 69, The stages - the seedling, the white bloom, the red bloom, the green bolls, the cracked boll and the open boll.

INSECT CONTROL

PAGE 67, Now Jack Shannon has very modern and expensive ag planes costing $650,000 or more. New planes have a capacity of 620 gallons. The Stearman's held as much as 80 to 100 gallons. They started in 1950 with a powder, calcium arsenate for the control of the boll weevil?

PAGE 70, Top Left: Cotton grower, Snookie Spaggiari of Clarksdale, Mississippi, opens a five-gallon can of Sweetkil, Sevin and molasses which attracted the moths of the bollworms and bud worms, while ground spray operator, Eddie Ceranti looks on. Bottom Left: One of the new experimental materials brought on the market for the control of worms was "Hot Sauce". Hopefully the worms would ingest the Hot Sauce when in contact with the leaves and/or fruit. Right: No doubt, this kind of application caused ag pilots to be called "crop dusters". Here in the late forties, in an effort to control the boll weevil, this pilot is dispensing calcium arsenate, which according to the "old timers", only made the boll weevil fat.

PAGE 71, Top Left: In the late forties, an ag pilot in his Stearman applies an insecticide for the control of thrips in young cotton. Top Right: A wide track ground sprayer applies insecticides to skip-row cotton, such as this two and one. Bottom: Here Eddie Ceranti applies an insecticide to eight rows of cotton in a four and four skip-row configuration. June-planted soybeans are in the four-row skip. The spray rig is home made.

PAGE 72, Left: With the Boll Weevil-Eradication Program in place many young cotton growers, especially in the Southeast where the program started, have never seen a boll weevil. Right: The cotton bollworm, if not controlled, will destroy blooms, small and large bolls. Although Bt cotton is active against both the tobacco bud worm and the cotton boll worm, the boll worm is less susceptible to the Bt toxin than the bud worm.

PAGE 73, Top Left: In 1957, consultants-entomologists are as essential for insect control as the insecticides and the machines for applying them. Pictured is Harris Barnes, III, Mississippi. Top Right: Bill Pellum, Mississippi Bottom Left: In 1957 Dr. Ivan Miles & Jack Oakman. Bottom Right: Charles Youngker and Al Linguel of Arizona.

PAGE 74, The mark of an excellent ag pilot, in the "good ole days" was one who was always "spinning his wheels" in the tops of the cotton plants. One such pilot was "Cotton" Carnahan, Clarksdale, Mississippi, shown here getting his Stearman cranked and ready for the next job.

PAGE 75, Left: Applying an insecticide with CDA's (controlled droplet applicators) in North Alabama. Right: Insecticides can be easily applied from a cultivator as seen here for the control of thrips, cutworms, etc.

PAGE 76, Left: The "needle nosed" aircraft of the nineties. Right: The helicopter has been very beneficial to the cotton farmer, especially with its capability of being able to apply insecticides and defoliants to small, irregular fields that are bordered by woods.

MATURE COTTON

PAGE 77, Top Left: In 1964 urban sprawl was taking over good farmland near Clarksdale, Mississippi. Notice the old barns that were once part of a farm headquarters. At the turn of the century all of this farmland had been "planted" to homes, apartment complexes and a private school. Top Right: In the South Mississippi Delta, this pretty white church stands out in a cotton field with a potential of two bales plus per acre. Bottom Left: In North Carolina many farmers have "turned their backs" on flue-cured tobacco and embraced cotton in a big way. PHOTO 4 - Many family cemeteries dot cotton fields through out the Mid-South.

PAGE 78, Left: Son, Dustin, passes a jug to his father, Larry Nelson, on a tractor at the Nelson Farm near Seagraves, Texas. Top Right: In August 1973 these growers were participating in an Extension Service sponsored field tour, designed to show how varieties, herbicides, insecticides, and fertilizers had performed during the year. Bottom Right: Larkin Martin of Courtland, Alabama, has climbed the ladder on Martin Farm, starting as an insect scout and working now as managing partner on the large North Alabama farm. She is also very active in the management of Servico, one of the first high-capacity gins in the Mid-South.

PAGE 79, Left: The Shaw's of North Alabama, near Tanner, "mark" their cotton rows early in the season to insure proper entry of each field with cultivators, spray machines, and cotton pickers. Right: Excellent cotton on the John Howard Farm near Deep Run, North Carolina.

PAGE 80, The Harwell home, The Columns, overlooks an excellent cotton field in full bloom near Florence, South Carolina.

PAGE 81, Left: In a cotton field you cannot ask for any better uniformity than this. Right: Beautiful cotton just outside of Yuma, Arizona. Note baled alfalfa hay and orchard in the background. Here temperatures get up to 100 plus degrees many days during the growing season.

OPEN COTTON FIELDS

PAGE 82, This excellent cotton field on the Freddie Ross farm, Scott, Mississippi, borders the origin of Deer Creek that winds through the South Delta of Mississippi. The land contiguous to this creek is very desirable and expensive.

PAGE 83, Left: Outside rows sometimes yield twice as much cotton as the interior row grown next to it. This is one of the reasons two and one skip-row cotton is so popular - all of the rows are outside rows.
Right: Open cotton in the San Joaquin Valley of California. Note mountains on the west side of the valley.

PAGE 84, Left: In the hot, dry summer of 2000, Deer Creek in the South Delta of Mississippi, almost went dry. Right: Cotton has come back strong in Georgia, now the largest cotton producing state in the Southeast. Here Bill Craven, Extension Leader of Burke County, one of the largest cotton-producing counties of the state, inspects a good well-defoliated crop on the farm of Percy Dixon.

PAGE 85, Left: Jim Hanson and sons, Philip and Erik, stand in three-bale plus cotton near Corcoran, California. Right: Sunset on a great top crop of cotton.

PAGE 86, Top Left: Stan Andrew of Gila River Farms, located near Sacaton, Arizona, where four and one skip-row cotton has been grown for many years. The defoliation has been done with ground machines and the harvest with two and four-row pickers. The farm belongs to the Gila River Indian Community. Top Right: A "Cabin in the Cotton" located on the Dudley Pillow Farm near Yazoo City, Mississippi. Bottom: Mark Borba, Borba Farms, Inc., Riverdale, California, grew the hybrid cotton, Chembred CB7, for several years.

PAGE 87, Top Left: Not a lazy bone in his body Larry Fenster of Fitzgerald, Georgia works as a Senior Captain for Delta Airlines four days a week and grows a cotton crop in his spare time. Top Right: Terry White of Altus, Oklahoma usually makes about 900 pounds of lint on irrigated land, only 250 pounds of lint on dry land. Bottom Left: Ann Ruscoe of Coahoma County, Mississippi was the first female County Extension Agent in the state. Bottom Right: One hundred percent cotton is Ms Cindy McDowell Bolt, Greenville, South Carolina, Miss Southern 500. She attended school in Bishopville, South Carolina and graduated in Horticulture from Clemson University.

PAGE 88, Growing cotton was a beautiful thing in West Tennessee in 2001.

DEFOLIATION

PAGE 89, Left: One of the best defoliant materials ever used in a cotton field was Aero Cyanamid's, "Black Annie" as 16-percent nitrogen fertilizer in a powder form. There were at least two schools of thought concerning the application-in the late afternoon when the material would diffuse throughout the plant, or in the morning dew, at which time the water positively activated it. Right: In the early sixties, a Stearman ag plane applies a liquid defoliant, either Def or Folex, on a mature field of cotton on the King and Anderson Plantation, near Clarksdale, Mississippi.

PAGE 90, Left: A high clearance ground sprayer applies a defoliant material to a good field of mature cotton near Casa Grande, Arizona. Right: Ronnie Everidge applies a defoliant to a cotton field on his farm near Pinehurst, Georgia. What happened to the photographer?

PAGE 91, Defoliation at sunset.

PAGE 92, Left: Johnny Lawrence of Elaine, Arkansas is a typical hard-working and intelligent Mid-South Farm Manager. Right: An ag plane applies a liquid defoliant to a cotton field in the Delta, noted for its wonderful flat land sunsets.

HAND/MECHANICAL PICKING

PAGE 93, Top Left and Right: Excellent grades of cotton were obtained through hand picking in the forties and fifties. Bottom Left: In the late forties, James Johnson and his family were out picking their own crop on Baugh Plantation, Sherard, Mississippi. Bottom Right: Even though there were a few mechanical pickers here and there, hand picking continued. In the 1950s good stands of cotton and weed control were still much of a problem for some growers.

PAGE 94, Left: Here the six and nine-foot pick sacks are brought to the weigh horse, the weights tabulated, and the 60 to 90 pound sacks "shook out" by a man standing on a 2 x 12 board. These two-bale Colby trailers were among the first with rubber tires. Right: Unloading a bale from a tenant cotton pen, with a No. 3 wash tub, into a two-bale wooden trailer at the Wheeler/Graham Plantation, near Coahoma, Mississippi.

PAGE 95, Top Left: The International Harvester one-row cotton picker made its appearance in the Mid-South in 1948. It was field-proven on the Hopson Plantation, Clarksdale, Mississippi. Low or high-drum cotton pickers were mounted on International H or M tractors that could be used in the fields during the growing season. Top Right: The mechanical picker gradually took over from the hand picker. The mechanical pickers came along at a very opportune time for much of the work force on the farms was moving to the industrial North, where the living was easier and the pay higher. Bottom: The owner was so glad to get rid of his two, one-row pickers that he did not dump the seed cotton from them nor finish out the rows.

PAGE 96, Top Left: A John Deere, two-row picker in good cotton-on the Harold Simmons Farm, Clarksdale, Mississippi. Right: The best field of cotton I have ever seen (four bales plus) was located on Gila River Farms, Sacaton, Arizona. The variety was Deltapine 41 planted in the four and one skip-row pattern. Bottom Left: Two, two-row Allis-Chalmers cotton pickers, a high-drum on the left and a low drum on the right, work side by side in the Mid-South.

PAGE 97, Prior to the invention of the module builder, "cotton kids" had a good place to play on open trailers. John McKinney of *Progressive Farmer* composed this picture for a cover. He used four "cotton kids" in 100-percent cotton, three trailers of different colors and heights, and one each of all the cotton pickers in production, at the time- the International Harvester, the John Deere, and the Allis Chalmers. The "cotton kids" are (left to right) Marne Anderson, Jim Barnes, Monty Dodson and Ms Jamye Barnes Lane.

PAGE 98, Top Left: Rob Lewis, Coahoma County Extension Agent and Leon Bramlett, owner, discuss his cotton harvest. Top Right: In the early sixties, Nita Abraham Ross shows off good machine picked cotton while Eddie Ceranti looks on. Bottom: Before the advent of the module builder, seed cotton was unloaded into trailers, large and small.

PAGE 99, Top Left: Bill Fields points out that in some wet spots all the picker needs is a little "push". Top Right: In the fall and winter, Mid-South cotton growers usually dual up and break out the four-wheel drives to be able to pick the remaining cotton from the fields. Bottom: Air ducts located near the bottom of the drum, cause the seed cotton to float in the air, enabling the spindles to get at least "one more whack" at the locks.

PAGE 100, Mike Moore, Waddell, Arizona, formerly produced brown, green and red cotton lint. To the right are fast color fabrics woven by Sally Fox.

PAGE 101, Top Left: Harvest on Pima Cotton in the Harquahala Valley, Arizona. Top Right: Excellent defoliated cotton in South Delta, Mississippi. Bottom: Joe B. Logan, Seagraves, Texas, produced this beautiful field of cotton (note stripper in the background) that survived five hailstorms and produced better than two bales of irrigated cotton per acre.

PAGE 102, Left: In 1989, a four-row Case IH picker harvests the bottom part of an Arizona cotton crop that had been defoliated earlier. Right: Last rows of the harvest season, a five-row picker harvests 30" rows on the Mitchener Farm, Sumner, Mississippi

PAGE 103, Top Left: The world's widest skip-row picker put together by Jimmy Hargett of Bells, Tennessee and owned and operated by Hood Harris of Courtland, Alabama. Top Right: Seven of the 2001 models of the John Deere six-row cotton picker on the Omega Plantation near Clarksdale, Mississippi. The outlay for these seven pickers was about $2 million. Bottom: In 2001 the Case IH six-row cotton picker was introduced commercially, however several cotton growers had assembled their own in previous years.

PAGE 104, Left: GPS Specialist, Tasha Wells, Tifton, Georgia, discusses the yield monitor on the cotton picker with Ernie Iler, Farm Manager, Pineland Plantation, Newton. Right: Lori Johnson Franklin operated a 10-row, 30-inch planter and a five-row cotton picker on Hood Farms, Perthshire, Mississippi for seven years. Now that she knows cotton so well she has moved up to a crop insurance adjuster.

PAGE 105, It does not get any better than this - four new John Deere six-row pickers eat up a field of two-bale plus cotton on Wildy Farms near Manila, Arkansas.

HUMAN INTEREST

PAGE 106, Left: Happiness is a "broken in" pipe full of Prince Albert tobacco. Top and Bottom Right: In the late fifties and sixties Hays House Movers helped "clear" many a house from cotton acres across the Mid-South.

PAGE 107, Top Left: A rear mounted, six-row cultivator works in a field that is flanked by some new homes built for the workers. The homes, built by the owners at a cost of $12,000 to $15,000 each, replaced the tenant houses. Bottom Left: A man on LP tank that supplies fuel for his new home. No more wood to tote! Right: Rolling a cigarette with Bull Durham at a Good Gulf station in central Alabama.

MODULES

PAGE 108, Left: Improved cotton crops and mechanical pickers caused real problems in the sixties. Center: In the late fifties and early sixties gin yards were so swamped with trailers full of cotton that cotton growers were buying new trailers, thinking about building new gins and passing a fifth of whiskey to the ginner to get a trailer or two ginned out of order. A number dumped seed cotton on the ground or blew it into old sheds or barns. Right: A two-row cotton picker dumps directly on the clean ground.

PAGE 109, Left and Center: Seed cotton from Irvin Plantation, King and Anderson Inc., is pulled by a fan directly from a truck and blown into an old empty shed. Sometimes a regular gin fan was used to unload large trailers and discharge the seed cotton into an empty barn Right: And of course, when the ginner caught up with the cotton on the gin yard it was a slow expensive job to load the seed cotton back into trailers.

PAGE 110, Left and Right: In an effort to reduce the crowded situations at the gins, and the purchase of new trailers, many growers in the Mid-South, Texas and the Farm West opted to dump their seed cotton into ricking devices stationed on the turn roads. Since there were no covers for the ricks, the seed cotton was exposed to the weather and all kinds of fire hazards. Zinn McNeil, University of Arkansas, Milton Smith, Texas Tech, and J.K. "Farmer" Jones of Cotton Incorporated, each played important parts in the research and development of these ricking devices.

PAGE 111, Left: Just behind the cotton gin and mechanical cotton picker in importance to the cotton harvest was the development of the module builder. It was sponsored with grower dollars through Cotton Incorporated in the early seventies. It was spearheaded by Lambert Wilkes and Jerome Sorenson of Texas A & M University and J. K. "Farmer" Jones of Cotton Incorporated. It not only meant a great savings to the cotton growers in the building of new gins and the buying of new trailers but it also improved the grades of the lint and eased the ginning of the seed cotton. Right: In 1972 cotton growers were amazed at the ease this transporter was able to load the module on a flexible steel mat for an easy transport to the gin.

PAGE 112, The end of the day.

PAGE 113, Left: Larry McClendon of Marianna, Arkansas, lays twine across the module to be able to secure the module cover from both sides. Right: A good harvest, plus a module builder and the know-how to build a well packed module of evenly distributed seed cotton makes for the "school solution".

PAGE 114 - Left: As modules are brought in from the field to the Gilliard Gin in Lake Providence, La., they are "calved" on to a special railroad flat car. These flat cars are loaded with sufficient modules to keep the gin running all night. Right: Modules in California waiting to be transported to the gin.

PAGE 115, Left and Right: Neat rows of modules in Arizona (left) and Mississippi (right) wait their turn to be transported to the gin.

PAGE 116, Left: Jaby Denton, Marks, Mississippi, passes along patriotic and religious messages with his modules spotted along a state highway near Marks, Mississippi. Right: In 2001 a "Golden Retriever" at the Bobo-Moseley Gin picked up a total of 195 bales from seed cotton that was left on the ground after the modules were moved to the gin.

GINS

PAGE 117, Left: A two-wheel, one-bale trailer is being pulled down a modern highway with a tractor equipped with 55-gallon drums up front for spraying during the growing season. Right: An old conventional gin with front sheds to protect trailers of seed cotton and to house the suck pipes used to pull the seed cotton into the gin.

PAGE 118, Left: A gin yard full of trailers to be ginned off. The overhead seed house is for easy unloading. Right: One of the first gins with "Disney World Colors" was built in the mid-sixties on King and Anderson Inc., Clarksdale, Mississippi.

PAGE 119, During the 2001-2002 ginning season at the Bobo-Moseley Gin, Lyon, Mississippi there were 1,142 modules on the peak day on the gin yards. There were 80,000 bales tied out, a record for this four-stand gin.

PAGE 120, In Arizona, where there is little or no rainfall during the harvest season, this gin in the Harquahala Valley blows seed into piles on the outside. Note the neat universal density bales and the modules in the background.

PAGE 121, Left: A new roller gin being built in Arizona. Roller gins are designed to separate the seed from the lint of long staple cottons like Pima. Right: Larry McClendon of Marianna, Arkansas built a three-stand, new gin in 1993 at the cost of $3.5 million. It is capable of putting out 45-50 bales per hour. It is now called the McClendon-Mann-Fenton Gin.

PAGE 122, The Boe Adams Gin at Leachville, Arkansas, is the world's largest with ten stands, (five on each side) produces 2,200 bales in 22 hours - 100 bales per hour or ten bales per gin stand. It was built in 1992 at a cost $12 million.

BALES /WINTER

PAGE 123, The Trail of the Ugly Bale - Top Left: First sampled at the gin and brought to the compress. Top Right: Second sampled at compress (flat bale - 55" long, 26" wide, 36"- 48" thick). Bottom Left: Compressed bale Bottom Right: "Away to domestic or foreign mills we go."

PAGE 124, Left: A universal density (UD) press. 55" long - 20"-21" wide - 24"-30" thick. Right: Neat bales of Pima cotton await shipment on the gin yard of the J.G. Boswell, Corcoran, California.

PAGE 125, Snow on stalks that have been cut.

PAGE 126, Wintertime in the Mid-South.

PAGE 127, Left: What do cotton farmers do during the winter? Attend auctions, meetings, and hunt deer, ducks and rabbits, etc. Right: Beauty – snow on an unpicked field.

PAGE 128, Pairs of mallards on a Mid-South Bayou.

PEOPLE

PAGE 129, Top Left: Chauncey Taylor, Morgan, Georgia markets "Little Bales of Cotton". Top Right: A tour group from cotton producing provinces of the Peoples Republic of China toured major growing areas of US in 1980. Bottom: James Lee Adams, Camilla, Georgia, was one of the first cotton growers to use a computer and alligators in his farming operation.

PAGE 130, Top Left: My first boss, Bill Connell and his twin brother, Willis of Connell Brothers in Clarksdale, Mississippi, sold Delta cotton all over the world. Top Right: When Billy Dunavant, Cotton Merchant of Memphis, speaks – cotton growers listen! Bottom: Officers of the Joseph Walker and Company, Cotton Merchants, Columbia, South Carolina, gather about Michelle Pitcher, 1985 Maid of Cotton, as she "pulls" cotton in their classing room.

CYCLE ON SUNDIAL

Chemicals, machinery, and practices for cotton production may change from year to year but an open boll of cotton is forever!